PAPERBACK EDITION PRINTED 2008

© Aladdin Books Ltd 2006

Designed and produced by
Aladdin Books Ltd
2/3 Fitzroy Mews
London W1T 6DF

First published in 2006 by
Franklin Watts
338 Euston Road
London NW1 3BH

Franklin Watts Australia
Level 17/207 Kent Street
Sydney NSW 2000

Franklin Watts is a division of
Hachette Children's Books

A catalogue record for this
book is available from the
British Library.

Dewey Classification: 333.792'4

ISBN 978 0 7496 8155 5

Printed in Malaysia

Editor:
Katie Harker

Design: Simon Morse
Flick, Book Design and Graphics

Consultants:
Jackie Holderness – former Senior Lecturer
in Primary Education, Westminster Institute,
Oxford Brookes University

Rob Bowden – education consultant,
author and photographer specialising
in social and environmental issues

Illustrations: Simon Morse

Picture researcher: Alexa Brown

Photocredits:
*l-left, r-right, b-bottom, t-top, c-centre,
m-middle.* Front cover, Back cover, 1, 2-3,
3mtl, 5tl, 5bl, 5br, 6tr, 7tr, 10tr, 13tr, 13b,
17tl, 17br, 19bl, 23br, 26tl, 27br, 29br, 30tr,
30br, 31bl – www.istockphoto.com. 3tl, 9tl,
12tr, 28tr – Corbis. 3mbl, 3bl, 5ml, 5tr, 8tl,
8bl, 10bl, 11tr, 18bl, 20tr, 20bl, 21tr, 22tl,
22bl, 23tl, 24tr, 24bl, 25tr, 25bl, 30mr, 31tl –
US Department of Energy. 4tl, 16tl, 21bl –
Digital Vision. 4ml, 16ml – Comstock. 4bl,
9mr, 16bl – John Simmons © The Geological
Society of London, www.geolsoc.org.uk. 6bl –
Brand X Pictures. 12bl, 29tr – NASA. 14tl –
Photodisc. 14bl – Hiroshima Peace Memorial
Museum. 15tl – United Nations Treaty
Collection. 15br – Global Peace Solution. 19tr
– FWS. 26br – Flat Earth. 27tl – Lawrence
Livermore National Laboratory.

Our World

Nuclear Power

By Sarah Levete

Aladdin/Watts

London • Sydney

CONTENTS

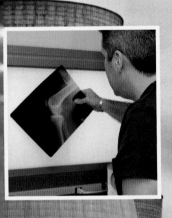

Notes to parents and teachers

This series has been developed for group use in the classroom as well as for children reading on their own. In particular, its differentiated text allows children of mixed abilities to enjoy reading about the same topic. The larger size text (A, below) offers apprentice readers a simplified text. This simplified text is used in the introduction to each chapter and in the picture captions. This font is part of the © Sassoon family of fonts recommended by the National Literacy Early Years Strategy document for maximum legibility. The smaller size text (B, below) offers a more challenging read for older or more able readers.

The advantages of nuclear power

Nuclear power is made without burning fuels that release harmful gases.

A

Coal

Uranium

◀ 1 kg of uranium produces as much energy as 3,000 tonnes of coal!

When the nucleus of some materials – such as uranium – is broken up, a huge amount of energy is released.

B

Questions, key words and glossary

Each spread ends with a question which parents and teachers can use to discuss and develop further ideas and concepts. Further questions are provided in a quiz on page 30. A reduced version of pages 30 and 31 is shown below. The illustrated 'Key words' section is provided as a revision tool, particularly for apprentice readers, in order to help with spelling, writing and guided reading as part of the literacy hour. The glossary is for more able or older readers. In addition to the glossary's role as a reference aid, it is also designed to reinforce new vocabulary and provide a tool for further discussion and revision. When glossary terms first appear in the text, they are highlighted in bold.

 ## See how much you know!

Which fuels can be burnt to produce energy?

Where is uranium found and how is it processed?

Why is there a shield in a nuclear power station?

How is nuclear energy used in medicine?

Which poisonous material is made from the nuclear fuel process?

How is nuclear energy used in space?

At the moment, which method is most commonly used to make nuclear energy: fission or fusion?

Key words

Radioactive

A

Atoms

Fuel   **Nuclear**

Plutonium   **Power**

Radiation   **Reactor**

Electricity

Glossary

Enrich – To improve or increase the nature of something.

Fission – The process of splitting an atom's nucleus into two, to release energy.

Fossil fuels – Fuels that form from the fossilised remains of prehistoric plants and animals. Coal, oil and gas are fossil fuels.

Fuel rods – The rods that are placed in a nuclear reactor to cause fission. Fuel rods are filled with uranium.

Fusion – The process of joining two atoms' nuclei together, to release energy.

B

Geiger counter – A device used to detect nuclear radiation.

Global warming – The warming of the Earth's atmosphere caused by gases trapping heat from the Sun.

Meltdown – When a nuclear reactor overheats so much that it starts to melt.

What is nuclear power?

Nuclear power is a type of energy. Some countries use it to make electricity to work lights and to power submarines. Nuclear power can also be used to make weapons and medicines. The energy is produced in a power station.

◀ **Energy makes things work.**

Energy powers machines. There are many types of energy. Nuclear energy comes from certain types of atoms. When we burn fuels such as coal, gas and oil we produce heat energy. This can be used to make electricity. The Sun's energy makes plants grow.

▶ We are surrounded by atoms.

Substances in the world around us, such as air and water, are made from tiny atoms. These can only be seen through very powerful microscopes. The main part of an atom is called a nucleus.

Inside an atom

- Electron
- Nucleus
- Neutron
- Proton

Nuclear energy comes from the force within an atom.

Each nucleus contains smaller parts called protons and neutrons. A nucleus is also surrounded by electrons. An incredibly strong force holds a nucleus together. When a neutron is used to split a nucleus into two, huge amounts of energy are released. This is called nuclear **fission**.

Nuclear fission

- Split nucleus
- Freed neutrons
- Proton
- Neutron
- Atom nucleus
- Energy released

 Can you name two things made of atoms?

All about uranium

The fuel used to produce nuclear power is called uranium. It is found in rocks in the ground. Uranium is a heavy silvery metal. It can be mined and made into nuclear fuel. The atoms in some types of uranium are easier to break up than some other substances.

◀ **Uranium is processed in a factory.**

Uranium is found deep in the ground. Miners dig out and crush rocks containing uranium. They use an acid to separate the uranium from the rest of the rock. The uranium is then turned into a gas and then into a powder. These processes **enrich** the uranium so that it can be turned into nuclear fuel.

◀ Uranium came from exploding stars!

Scientists believe uranium was formed when old stars exploded over six billion years ago. The uranium dust from the stars scattered throughout the Universe, eventually settling in rocks deep inside planet Earth.

Uranium is radioactive.

A radioactive substance gives off invisible rays of energy called radiation. These rays cannot be seen, tasted or felt. Uranium's radioactivity provides some of the heat inside the Earth. This is spread out over huge areas and is harmless. However, radioactivity from nuclear fuel is more concentrated and can be very dangerous.

Special machines called Geiger counters check the level of radiation.

 Why is uranium used as a nuclear fuel?

Nuclear energy

Today, nuclear power is mainly used to make electricity. This is usually cheaper than the electricity made from burning fuels such as coal, oil or gas. Some people say that nuclear energy is cleaner because it does not produce smoke.

◀ These stored uranium fuel rods can be used to make electricity.

A nuclear reactor is a large tank or building inside the power station. Here, pieces of powdered uranium are put into special **fuel rods**. These are gathered into bundles and placed in the centre of the reactor. The fuel rods make the atoms split to release energy that is then used to heat water.

▶ Fresh fuel is added to a nuclear reactor about every 18 months.

When the uranium inside the fuel rods loses its energy, the fuel rods are sent to special reprocessing stations. Here, any unused uranium is taken out so that it can be made into nuclear fuel again.

The heat energy that comes from splitting uranium atoms is used to make steam.

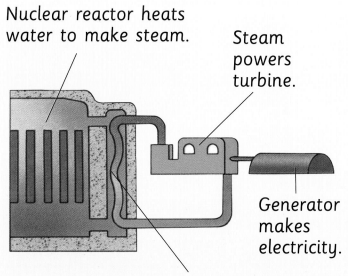

Nuclear reactor heats water to make steam.

Steam powers turbine.

Generator makes electricity.

Heat is transferred to other water.

This steam spins a turbine to drive a generator, producing electricity. In an oil-, gas- or coal-fired power station, fossil fuels are burnt to create heat to make steam. The same principle applies in a nuclear power station.

 How do you use electricity at home?

Other uses of nuclear energy

Nuclear energy has many other uses. Small amounts of radioactive materials can detect illnesses and cure some diseases. Some hospital equipment is also kept especially clean or sterilised with radiation.

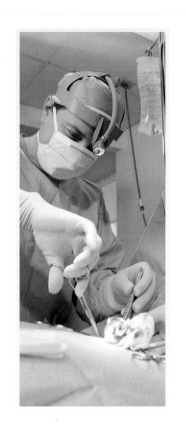

◀ **Radiation can be used to kill germs in food.**

Radiation can be passed through food to kill germs. This makes the food last longer. Astronauts eat radiated food to prevent them from becoming unwell in space. However, some people argue that because we don't know the effects of radiation on the body, it is unhealthy to radiate food.

▶ Radiation can be used to find and treat diseases.

Radioactive chemicals help to show up problems in a person's body. Radiation can also be used to treat diseases such as cancer, by destroying the harmful cells. X-rays that show up tissues and broken bones are also a type of radiation.

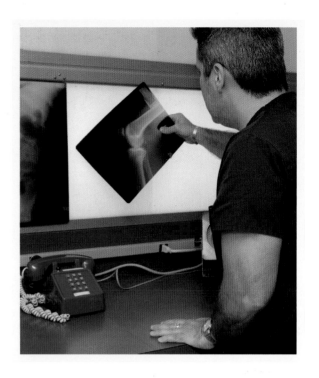

A nuclear submarine can stay at sea for months.

One tennis ball-sized lump of uranium can produce enough energy to power a nuclear submarine! The submarine can stay underwater for long periods of time without having to surface to collect extra supplies of fuel.

 Where do you think nuclear power stations should be built?

Nuclear weapons

The most terrible and destructive weapons are made from nuclear energy. These are nuclear or atomic bombs. So far, nuclear bombs have only been used a few times but they have caused the deaths and illnesses of hundreds of thousands of people.

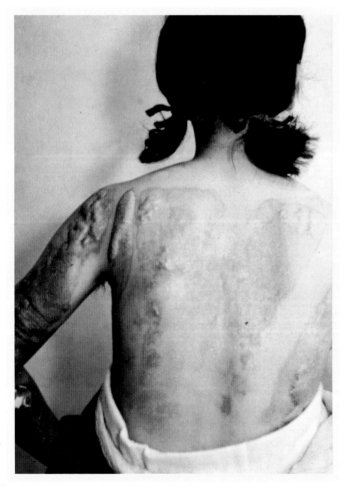

◀ **This woman is suffering from radiation from a nuclear bomb.**

During the Second World War, American pilots dropped a nuclear bomb on the Japanese city of Hiroshima. It killed 100,000 people. Another bomb killed 75,000 people in the city of Nagasaki. People in these cities continue to suffer from illnesses caused by the radiation.

Many leaders have agreed not to use nuclear power to make weapons.

Some people think that nuclear weapons should be destroyed to stop people using them. Others argue that owning nuclear weapons prevents wars, because people will be afraid of the weapons being used.

Nuclear weapons contain deadly plutonium.

When uranium is used to make nuclear energy, a material called plutonium is produced. This is one of the most toxic substances known to humans. Plutonium has been used to make nuclear weapons and can destroy a population within minutes.

 Why are nuclear weapons so dangerous?

The advantages of nuclear power

Nuclear power is made without burning fuels that release harmful gases. Some people argue that nuclear power is also cheap because only small amounts of fuel are needed to create lots of energy.

Coal

Uranium

◀ **1 kg of uranium produces as much energy as 3,000 tonnes of coal!**

When the nucleus of some materials – such as uranium – is broken up, a huge amount of energy is released. Nuclear power stations only use a small amount of fuel, unlike the large quantities of coal, gas and oil needed to make electricity. This could make nuclear power a cheap source of alternative energy.

◀ **Once a nuclear reactor has been built, it can make cheap electricity.**

We are using more and more energy but we are running out of fuels that make electricity, such as coal, gas and oil. As they run out, these fuels are becoming more expensive. Nuclear reactors are very expensive to build, but they only use a small amount of fuel.

Nuclear power does not release harmful gases.

When we burn **fossil fuels** we release harmful gases into the air. The gases trap heat around our planet, which is causing **global warming**. Nuclear power does not produce this type of pollution.

 Why do we need to find other energy sources?

The disadvantages of nuclear power

Many people worry about the safety of nuclear energy and the radiation it produces. There are also concerns about how to safely throw away or reuse radioactive waste material.

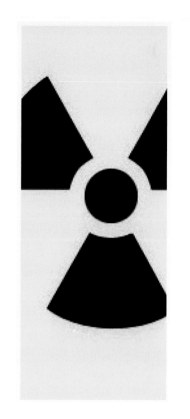

◀ Workers in a nuclear power station need to wear special clothing.

Radiation can make people very ill. Strict safety procedures need to be used in a nuclear power station. Special clothing protects workers from the radiation. Many people believe that radioactive materials cause disease and disabilities in people and animals living nearby.

▶ Mining for uranium can spoil the natural landscape.

A lot of rock has to be dug up to find small deposits of uranium. This can destroy the homes of animals and plants. Water can also become polluted during the mining process.

The materials used in nuclear fuel are deadly, and could be used to harm people.

Because nuclear materials are so dangerous, the security at nuclear power stations has to be very tight. If someone stole just a tiny amount of plutonium it could be used to cause terrible damage. When nuclear materials travel by truck, there is usually a special police convoy to protect them.

 Why is there tight security around a nuclear power station?

Nuclear accidents

Nuclear accidents have been caused by leaking pipes and equipment that has broken down. If radiation leaks, it can travel very far, harming people and the natural world. Power stations are now built to very high safety standards.

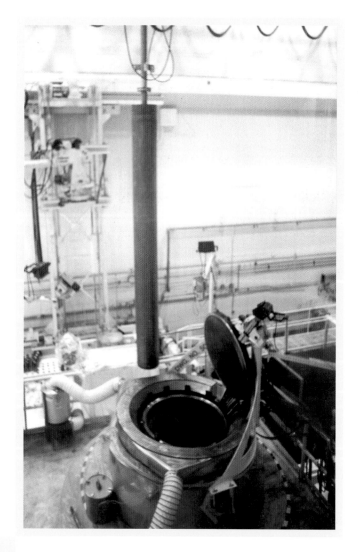

◄ **Nuclear reactors can overheat and burn.**

If a nuclear power station overheats, it can lead to a **meltdown**. This is when the nuclear reactor burns and becomes so hot that it melts. Equipment explodes and the heat is so intense that even concrete walls burn. A meltdown can allow deadly radioactive materials to escape, killing anything nearby.

▶ Human error caused the world's worst nuclear disaster at Chernobyl, Ukraine.

In 1986, mistakes by workers caused a meltdown at the Chernobyl nuclear reactor. An invisible cloud of poisonous radiation spread for many kilometres. Today, more advanced machinery can help prevent this kind of disaster.

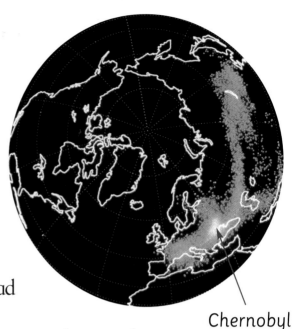

Chernobyl

Pink area shows spread of radiation.

Nuclear radiation can kill.

The powerful radiation from nuclear fuel can cause illness and death to people and wildlife. The children of people exposed to radiation may be born with illnesses and disabilities. However, it is not always easy to prove the link between radiation and ill health.

 Why do some people think that nuclear power is dangerous?

Nuclear power and safety

There are many safeguards to make sure that nuclear power stations stay safe and secure. Safety procedures are constantly checked. Special containers are also used to carry the nuclear fuel or waste to and from the power station.

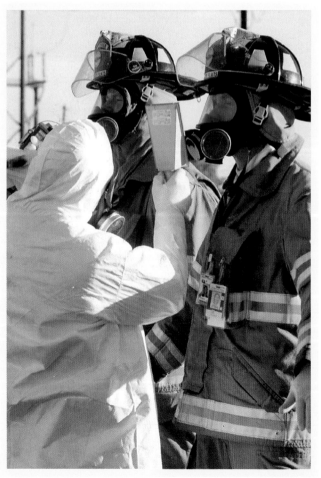

◀ **Special machines test for levels of radiation.**

Everyone who works in a nuclear power station or with nuclear fuel is regularly tested to check levels of radiation. Machines called **Geiger counters** can work out if the levels of radiation in a person's body are safe. Even clothes and masks are tested!

Transporting nuclear fuel is dangerous.

There are only a few countries where factories can reuse nuclear fuel rods. Old and used fuel rods are shipped by sea or driven long distances to be processed. The fuel rods must be carried in specially built containers that will not burn or crack.

Nuclear power stations are surrounded by thick walls.

Power stations cover hundreds of square metres. The reactor is surrounded by thick concrete and steel walls to prevent dangerous fuel leaks. But many people believe nuclear power stations will never be safe enough.

 How can nuclear accidents be prevented?

Nuclear waste

No one knows what to do with waste from a nuclear reactor. Nuclear power stations have to deal with materials that cannot be used again. The problem is that nuclear waste can remain radioactive for thousands of years.

◀ **Used fuel rods take years to cool down.**

When fuel rods cannot be used any longer, they are cooled in special ponds in the nuclear power station to allow their heat and radioactivity to decrease. It can take up to 50 years for them to cool down! After this time, they are chopped up and dissolved in acid so that they can be disposed of more safely.

▶ Dangerous nuclear waste can be stored in strong metal containers.

Some nuclear waste can be turned into a dry powder that is mixed with glass and poured into steel tubes. However, very few people want to live near a nuclear waste store.

Nuclear waste can be buried underground.

One option is to bury sealed nuclear waste underground. However, we do not know how this may affect the nearby land, crops and rivers. The US is planning to use the Nevada desert as a nuclear dumping ground but people living nearby are very worried.

How would you *feel* about living near a nuclear store?

Nuclear energy today

Accidents have made nuclear power unpopular. But we are in danger of running out of fossil fuels. Some people think that nuclear power is the most reliable alternative to coal, gas and oil. It is argued that nuclear power also causes less pollution than fossil fuels.

► **Nuclear power plants supply about 17% of the world's electricity.**

Some countries, such as France, depend on nuclear energy for most of their electrical supply. Other countries, such as the US, the UK and China, have mainly used coal and oil to produce their electricity. However, these countries are beginning to look again at nuclear energy as an option for the future.

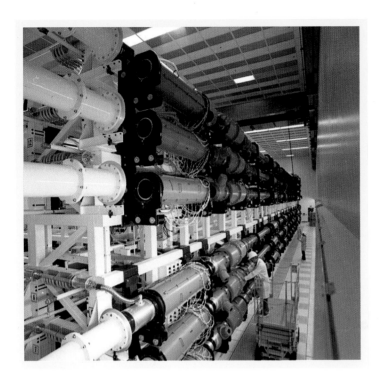

◀ New power stations are being developed.

Scientists are trying to make nuclear energy safer. Some nuclear reactors now use laser beams to produce a nuclear reaction. Small amounts of fuel are needed and the reaction only works under certain conditions.

Some countries are using other forms of energy.

Some countries, such as Germany, are using less nuclear power. They are now turning to other forms of energy, such as wind and solar power. These sources only produce small amounts of energy but they are safe, widely available and cause less pollution.

 What other sources of energy could we use?

Future developments

Nuclear power may become more popular in the future. Oil and gas will run out one day because developing countries are using more and more electricity. We must therefore find alternative energy sources.

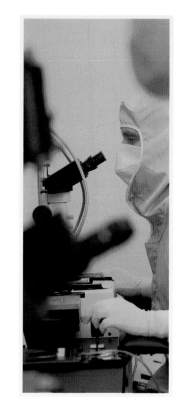

Nuclear fusion

◀ In 50 years' time, nuclear energy may be made in a different way.

Atoms

Energy released

Heavier atom

Most nuclear reactors use fission to make energy. Another way to create nuclear energy is to join atoms together. This is called **fusion**, and produces much less radiation than fission. Fusion takes place in special doughnut-shaped reactors called tokamaks, but so far scientists have not found a way of producing enough energy from fusion.

▶ Nuclear reactors could send spacecraft to other planets.

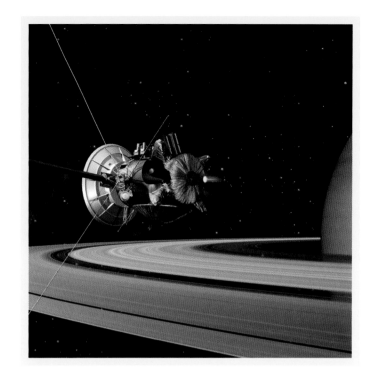

Small nuclear reactors could provide enough energy to power a spacecraft. However, people argue that it would be dangerous to launch nuclear materials into space.

Nuclear waste may be a problem for your children and their great-great-grandchildren.

Dumping nuclear waste may solve the problem for now, but how will future generations deal with the waste?

Scientists have suggested writing instructions on special paper that does not rot. This paper is similar to the papyrus used by Ancient Egyptians thousands of years ago.

 What do you think of nuclear power?

Which fuels can be burnt to produce energy?

Where is uranium found and how is it processed?

Why is there a shield in a nuclear power station?

How is nuclear energy used in medicine?

Which poisonous material is made from the nuclear fuel process?

How is nuclear energy used in space?

At the moment, which method is most commonly used to make nuclear energy: fission or fusion?

Key words

Radioactive

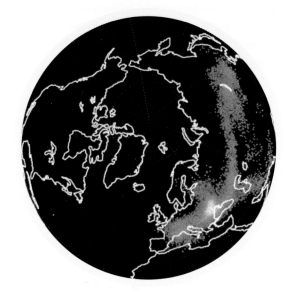

Atoms   **Energy**

Fuel   **Nuclear**

Plutonium   **Power**

Radiation   **Reactor**

Uranium

Electricity

Glossary

Enrich – To improve or increase the nature of something.

Fission – The process of splitting an atom's nucleus into two, to release energy.

Fossil fuels – Fuels that form from the fossilised remains of prehistoric plants and animals. Coal, oil and gas are fossil fuels.

Fuel rods – The rods that are placed in a nuclear reactor to cause fission. Fuel rods are filled with uranium.

Fusion – The process of joining two atoms' nuclei together, to release energy.

Geiger counter – A device used to detect nuclear radiation.

Global warming – The warming of the Earth's atmosphere caused by gases trapping heat from the Sun.

Meltdown – When a nuclear reactor overheats so much that it starts to melt.

Index